# Cambridge Elements ☰

Elements of Paleontology
edited by
Colin D. Sumrall
*University of Tennessee*

# CRINOID FEEDING STRATEGIES: NEW INSIGHTS FROM SUBSEA VIDEO AND TIME-LAPSE

David L. Meyer
*University of Cincinnati*

Margaret Veitch
*University of Michigan*

Charles G. Messing
*Nova Southeastern University*

Angela Stevenson
*University of British Columbia, Vancouver and GEOMAR*

Paleontological
SOCIETY

CAMBRIDGE
UNIVERSITY PRESS

# CAMBRIDGE
## UNIVERSITY PRESS

University Printing House, Cambridge CB2 8BS, United Kingdom

One Liberty Plaza, 20th Floor, New York, NY 10006, USA

477 Williamstown Road, Port Melbourne, VIC 3207, Australia

314–321, 3rd Floor, Plot 3, Splendor Forum, Jasola District Centre, New Delhi – 110025, India

79 Anson Road, #06–04/06, Singapore 079906

Cambridge University Press is part of the University of Cambridge.

It furthers the University's mission by disseminating knowledge in the pursuit of education, learning, and research at the highest international levels of excellence.

www.cambridge.org
Information on this title: www.cambridge.org/9781108810074
DOI: 10.1017/9781108893534

First published 2021

*A catalogue record for this publication is available from the British Library.*

ISBN 978-1-108-81007-4 Paperback
ISSN 2517-780X (online)
ISSN 2517-7796 (print)

Additional resources for this publication at www.cambridge.org/meyer-resources.

# Crinoid Feeding Strategies: New Insights from Subsea Video and Time-Lapse

Elements of Paleontology

DOI: 10.1017/9781108893534
First published online: February 2021

David L. Meyer
*University of Cincinnati*

Margaret Veitch
*University of Michigan*

Charles G. Messing
*Nova Southeastern University*

Angela Stevenson
*University of British Columbia, Vancouver and GEOMAR*

**Author for correspondence:** David L. Meyer, david.meyer@uc.edu

**Abstract:** Modern videography provides an ever-widening window into subsea echinoderm life with vast potential for new knowledge. Supported by video evidence throughout, this Element begins with time-lapse video made in 1983 on film, using an off-the-shelf camera, flash, and underwater housings. Although quality has now been significantly improved by digital imagery, films from over 30 years ago captured crinoid feeding behavior previously unknown and demonstrated a great potential to learn about many other aspects of their biology. This sequence is followed by several examples of recent digital videography from submersibles of deep-sea crinoids and remotely operated vehicles (ROVs) (stalked and unstalked), as well as close-up video of crinoids in aquaria. These recent studies enabled a new classification of crinoid arm postures, provided detailed views of food-particle capture, and revealed a wide range of behaviors in taxa never before seen in life.

**Keywords:** crinoid feeding strategies, echinodermata, crinoidea, feeding, recent, video

ISBNs: 9781108810074 (PB), 9781108893534 (OC)
ISSNs: 2517-780X (online), 2517-7796 (print)

# Contents

Please be aware that this title makes heavy use of video content.
In case of any playback issues, high quality versions of the
video files are available for download in the following location:
www.cambridge.org/meyer-resources.

# 1 Feather Stars at Lizard Island, Great Barrier Reef (14° 38' S, 145° 30'E)

## David L. Meyer

**Video 1** Day–night time-lapse record of a cluster of feather stars at 10 meter depth on a fringing reef, Lizard Island, Great Barrier Reef. Video available at www.cambridge.org/meyer-resources.

Video 1 is a time-lapse record of a cluster of feather stars at 10 meter depth on a fringing reef. It was filmed in April, 1983, using a Super8 film camera at 1 frame per minute, starting at 1535 hrs, continuing through the night and following day for a total of 25 hr (Meyer, 1997). During daylight, metered exposures were made with available light, and after dark, illuminated by an electronic flash. The current was unidirectional from the southeast and related to the tidal cycle. Current velocity increased to about 25 cm/s just after LW, slacking to 0 at HW, peaking again at about 10 cm/s during the ebb-tide.

A recording current meter placed near the crinoids shows that the crinoids form filtration fans at currents speeds below the 5 cm/s threshold sensitivity of the meter (Fig. 1; also Meyer, 1997). As current speed increases, the filtration fans are increasingly deflected but individuals identified as *Himerometra robustipinna* (P. H. Carpenter) maintained position even at maximum current speed > 20 cm/s. A nonfeeding posture with arms curled over the oral disk occurred only at dead slack water. Some individuals of *H. robustipinna* situated in close proximity form "common fans" of overlapping arms. In Meyer et al. (1984), some of these "partners" were identified as *H. bartschi* (A. H. Clark), but subsequent taxonomic revision (Taylor et al., 2017) has determined *H. bartschi* to be a junior synonym of *H. robustipinna*.

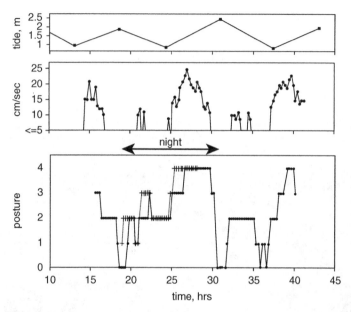

**Figure 1** Tide height, current velocity, and behavior of feather stars as shown in Video 1, Lizard Island, GBR. H. robustipinna =•, D. articulata =+. Feeding postures: 0 = no feeding fan; 1=partial fan, 2=full fan; 3=deflected fan; 4=extreme deflected fan From Meyer, 1997

Individuals of *H. robustipinna* responded to current speed fluctuations and formed filtration fans day and night. Shortly after nightfall, a single individual identified as *Dichrometra* (formerly *Liparometra*) *articulata* (Müller) crawled from the lower margin of the frame to a perch on a substrate ridge and formed a filtration fan. This crinoid did not move directly to its perch but first approached the ridge occupied by two individuals of *H. robustipinna*; it then moved a short distance away before forming its feeding fan. As current increased, this crinoid was strongly deflected until it disappeared from view and did not reappear when current abated. Other crinoid species may be present in this assemblage but are semi-cryptic, extending only the arms from a crevice.

## 2 Arm Postures in Living Crinoids

### Charles G. Messing

This section provides an overview of a newly proposed classification of arm postures for crinoids both living and extinct (Messing et al., in review) for the online, revised crinoid volume of the *Treatise on Invertebrate Paleontology*. The order of posture types in the video follows the submitted article, which includes many still images in color and black and white, as well as line drawings. The

**Video 2** Arm postures for living and fossil crinoids, a new classification. Video available at www.cambridge.org/meyer-resources

videos were gathered from a variety of sources and are mostly from deep water. Some of the crinoids featured have only recently been observed in the wild and photographed for the first time. Brief descriptions of each posture type, taxa, depth, location, and source follow for each clip.

1 Fan posture / monoplanar – "Antedonid" feather star on dead stalk.

2 Food-particle movement along a food groove – unidentified feather star. © Austin Eunice (YouTube).

3 Maximum food particle size difficult to pass (large particle feeding in an aquarium) – unidentified feather star. © aquabacs (YouTube).

4 Fan posture / monoplanar – "Antedonid" feather star on dead stalk, Line Islands (NOAA).

5 Fan posture / chiefly biplanar – *Crinometra brevipinna* (Charitometridae) meadow, Caribbean Sea (NOAA).

6 Fan posture / chiefly biplanar – *Crinometra brevipinna* row, Caribbean Sea (NOAA).

7 Fan posture / multiplanar – unidentified feather star, Marianas Islands (NOAA).

8 Fan posture / multiplanar – ?*Comaster schlegelii*, Apo Reef, Mindoro, Philippines (NOAA).
*Clarkcomanthus alternans*, Raja Ampat, Indonesia, © C G Messing.
*Clarkcomanthus alternans*, Lizard I., Australia, © D L Meyer.

9 Disk posture – "Antedonid" feather star on glass sponge, Caribbean Sea (NOAA).

10 Disk posture – Hyocrinidae, Indonesia (NOAA).

11  Parabolic posture – *Metacrinus* sp. (Isselicrinidae) downcurrent view, Western Pacific (NOAA).

12  Parabolic posture – *Metacrinus* sp. upcurrent view, Western Pacific (NOAA).

13  Parabolic posture – *Proisocrinus ruberrimus* (Proisocrinidae), Wake I. (NOAA).

14  Parabolic posture – *Phrynocrinus* sp. (Phrynocrinidae) (NOAA).

15  Parabolic posture – Some arms spread in response to stronger flow further from substrate – *Endoxocrinus carolinae* (Isselicrinidae), Isla Roatán, Honduras. ©Karl Stanley.

16  Parabolic posture – *Stylometra spinifera* (Thalassometridae) feather stars on *Bathypathes* antipatharian (NOAA).

17  Transition between parabolic and disk postures – Hyocrinidae, Indonesia (NOAA).

18  Conical feeding posture – *Guillecrinus neocaledonicus* (Guillecrinidae), Pacific Islands Marine National Monument (NOAA).

19  Conical feeding posture – *Guillecrinus neocaledonicus*, Pacific Islands Marine National Monument, closer view to show podia extended for feeding (NOAA).

20  Independent arm posture – *Comactinia meridionalis hartlaubi* (Comatulidae), Isla Roatán, Honduras, ~150 m. ©Messing/ Baumiller.

21  Multidirectional posture – *Davidaster rubiginosus* (Comatulidae), Curaçao. © NaturePicsFilms, Yvonne & Tilo Kühnast, www.naturepicsfilms.com.

22  Common fans – *Metacrinus* sp., Marianas Islands (NOAA).

23  Common fans – Charitometridae feather stars, Marianas Islands (NOAA).

24  Canted conical posture – *Holopus rangii* (Holopodidae) (NOAA).

25  Nonfeeding postures – Collecting bowl – Atelecrinidae, Caribbean Sea (NOAA).

26  Nonfeeding postures – Collecting bowl – unidentified feather stars, Line Islands (NOAA).

27  Nonfeeding postures – Collecting bowl – unidentified feather stars, Line Islands – CAVEAT! Close-up showing extended podia – these feather stars are actually feeding (NOAA).

28  Nonfeeding postures – Wilted flower – *Endoxocrinus carolinae*, Isla Roatán, Honduras. ©Karl Stanley.

29  Nonfeeding postures – Wilted flower – *Metacrinus* sp., Marianas Islands (NOAA).

## Stalked Crinoids Time-Lapse Series

- Platform: *Johnson Sea-Link I submersible*
- Photosea 1000A 35-mm U/W camera system
- Photosea 1500X strobe lights
- Deployed: *JSL I* dive 3577, 10 August 1993
- Recovered: *JSL I* dive 3634, 7 November 1993
- 165 frames at 5-hour intervals
- Variations in crinoid postures reflect changing flow velocity and direction.
- Variations in current direction due to meanders & eddies in the Florida Current.

**Video 3** Time series of *Neocrinus decorus* (Isocrinidae) at 5-hr intervals for 37 days, Bahamas, ~420 m. Video available at www.cambridge.org/meyer-resources

**Video 4** Particle interception and transport in a comatulid crinoid *Florometra serratissima*. Video available at www.cambridge.org/meyer-resources

### 3 Mechanism for Particle Interception and Transport in Comatulid Crinoid *Florometra serratissima*: Presenting a Range of Particle Sizes from Mesocosm Observations

## Angela Stevenson

Aquaria technology has advanced tremendously through time and inadvertently enabled fine-scale examination of food-particle interception and transport in crinoids. Video 4 was filmed with a Sony Digital HD Video Camera Recorder (HDR-SR12) and edited in iMovie. The comatulid crinoid *Florometra serratissima* was kept in captivity for six months in recirculating seawater aquaria bubbled constantly with ambient air and equipped with a multistage filtration system, including biological filter (sock filtration, protein skimmer, and bio-balls) and UV sterilizer. This long period of observation provided multiple opportunities to witness a wide range of feeding behaviors. In Video 4, three Fauna Marine Ultra min F food particles (in center of video frame) are rapidly transported to the central mouth of *F. serratissima*. The particles fit snuggly between the tertiary podia along the ambulacral margin of the food groove, ensuring their successful delivery to the central mouth for consumption. In this six-month mesocosm study, *F. serratissima* was most successful at capturing Ultra min F over other food types, which were comprised of a different range of particle sizes.

Comatulid functional morphology is designed to capture and transport a certain size of particles more effectively than others and this is limited primarily by the width of the ambulacral groove (Meyer, 1982; Baumiller, 1997). However, crinoids are nonselective suspension feeders that will attempt to feed on any particle intercepted by their podia (Meyer, 1982), including the eggs

**Video 5** Capture and transport of crinoid eggs. Video available at www.cam bridge.org/meyer-resources

**Video 6** Capture of mysid shrimp that exceed width of ambulacral groove. Video available at www.cambridge.org/meyer-resources

of a female crinoid. The unfertilized eggs of *F. serratissima* are perfectly spherical, tiny ($207 \pm 6$ μm; Mladenov and Chia, 1983), and have a high energy content (6.02 joules/mm$^3$; McEdward et al., 1988) – making them a desirable food source, even if mildly cannibalistic and challenging to capture. Video 5, filmed with a Sony Digital HD Video Camera Recorder (HDR-SR12) and edited in iMovie, displays numerous eggs (tiny white specks against the blue mesocosm wall) floating in the aquarium's seawater after a natural spawning event has occurred in the tank. Eggs flutter past the pinnules of *F. serratissima,* some come in contact with their primary podia but few are successfully flicked into or retained in the pinnule food groove. Nonetheless, near the end of the video, several eggs have accumulated in the arm food groove and are rapidly moving mouthward.

Frozen mysid shrimp provide a nutritious diet for fish and invertebrates, but their size range (0.3–1.8 cm) greatly exceeds that of an ambulacral food groove, making them an unlikely food source for crinoids. The challenge here is retaining these large particles in the food groove to triumphantly deliver a mysid feast to the central mouth of the crinoid. This first seems like an impossible task for

**Video 7** Time lapse of commensal crab next to crinoid. Video available at
www.cambridge.org/meyer-resources

*F. serratissima*, but, in Video 6, filmed with a Sony Digital HD Video Camera
Recorder (HDR-SR12), and edited in iMovie, several pieces of mysid shrimp
have been captured and are actively being transported to the central mouth. The
mechanism for large-particle interception and transport differs from that used for
the much smaller Ultra min F (Video 4) and eggs (Video 5). Typically, a food
particles is first directly intercepted by the primary and secondary podia of the
pinnules, then flicked into the pinnular food groove where it is controlled by small
tube feet (tertiaries) that move the particle in a coordinated manner between
adjacent podia toward the arm food groove (Lahaye and Jangoux, 1985;
Baumiller, 1997). The material is then swept rapidly toward the crinoid's
mouth via the arm food groove. In Video 6, the successful initial capture and
subsequent transport of the mysid shrimp is a coordinated effort between several
adjacent pinnules. Incapable of fitting into the relatively narrow food groove, the
mysid shrimp is enveloped by adjacent pinnules and is periodically seen bulging
out from between the pinnules holding it in place in the food groove. It is slowly
passed, in a similar synchronized manner, from one pinnule to the next down the
arm. It is unclear how the mysid shrimp is taken up once it reaches the central
mouth and if it is consumed.

Crinoids are passive suspension feeders and play an important role in bentho-
pelagic coupling. Night mode and time-lapse video footage allow for covert
recordings of the intimate relationship between crinoids and their symbionts,
like the decorator crab *Chorilia longipes* on comatulid *Florometra serratissima* in
Video 7, filmed with a Sony Digital HD Video Camera Recorder (HDR-SR12)
coupled with a Pclix XT Time-lapse Photography Controller, and edited in
iMovie. While *C. longipes* is never directly seen holding the crinoid's fecal
pellets between its claws, it spent the majority of the six months perched near
the anal papilla (seen in the background, just beyond *C. longipes*) of its host.

**Video 8** *Florometra serratissima* pentacrinoid postlarva. Video available at
www.cambridge.org/meyer-resources

## 4 Feeding Postures in a Pentacrinoid *Florometra* and Responses of *Democrinus* (Bourgeticrinidae) and *Cenocrinus* (Isocrinidae) to Increased Current

### Margaret Veitch

For Video 8, four hours of video was captured by a Hero5 GoPro attached to a diving weight positioned on the sea floor near Tuwanek, British Columbia, on June 8, 2019, at 40 meters. GoPro deployed by Stevenson/Veitch. The ~2.5 cm tall *Florometra serratissima* pentacrinoid postlarva kept its crown elevated and away from the hard substrate for the duration of the video (Fig. 2). Arms are outstretched, nearly perpendicular with tips lightly curled upward, creating a hook-like appearance. Previously reported feeding positions for adult *F. serratissima* include arms held in a cone posture in no current and creation of a partial fan posture in mild currents (Byrne and Fontaine, 1983). Neither position is observed during the hours of recorded video, suggesting that pentacrinoid postlarva might have different feeding postures than in their adult form.

Minor current is visible during Video 8 via movement of suspended particles upslope (~1 cm/s). The pentacrinid rotates the crown slowly during several three- to four-minute segments of the recordings, about four times across the four hours. The rotation does not seem to be in response to any visible changes in current. The crown appears to rotate both into and out of the visible current, without the aid of the arms as has been seen in crown rotation for stalked crinoids (Baumiller et al., 1991).

Video 9, *Democrinus* sp. captured from inside the submersible *Idabel*, using a handheld SONY video camera (recorded by Baumiller/Veitch), at 240 meter depth. Green laser scale: 10 cm. Feeding posture in 3 cm/s current. Arms are held conically (~ 60 degrees from mouth), while stalks adopt a range of slight to large bent postures oriented downcurrent in response to the current. Several

**Figure 2** *Florometra serratissima* pentacrinoid postlarva. Diameter of disk tag ~1.5 cm. Photo by Veitch.

**Video 9** *Democrinus sp.,* December 16, 2016, off Isle Roatan, Honduras. Video available at www.cambridge.org/meyer-resources

**Video 10** *Cenocrinus asterius*, May 16th, 2016, Isla Roatán, Honduras. Video captured from inside the submersible Idabel, using a handheld SONY video camera (recorded by Baumiller/Veitch). 150 meters depth

Video from submersible approaching *Cenocrinus asterius* from the aboral face. Crowns are held in a parabolic feeding position in low current (< 15 cm/s). After pulling out and back with the submersible, we again approach the pair of *C. asterius* from behind. In doing so, we generated a large push of current, around 25 cm/s. Both *C. asterius* lose their parabolic fan position as the current hits and continue to alter over next 20 seconds as secondary swells of currents occur. The feather star potioned between them has a much less drastic reaction, with only about the proximal 2/3 of its arm moving with the initial current hit, then returning back to its previous feeding position until the second current hit where it moves again slightly. Video available at www.cambridge.org/meyer-resources

**Video 11** *Cenocrinus asterius*, May 18, 2016, Isla Roatán, Honduras. Video available at www.cambridge.org/meyer-resources

individuals collapse from feeding position at around 7 cm/s. Stalk position does not change along with collapsing of the crown from feeding positions at 7 cm/s.

Video 10, *Cenocrinus*. Feeding posture in 3 cm/s current. Arms are held conically (~ 60 degrees from mouth), while stalks adopt a range of slight to large bent postures oriented downcurrent in response to the current. Several individuals collapse from feeding position at around 7 cm/s. Stalk position does not change along with collapsing of the crown from feeding positions at 7cm/s.

Video 11, captured from inside the submersible *Idabel*, using a handheld SONY video camera (recorded by Stevenson/Veitch), at 150 meters depth, shows a series of *Endoxocrinus* on sides and top edges of two limestone outcrops. Submersible is moving into a very weak current, ~2 cm/s. *Endoxocrinus* fans are oriented in multiple directions but mainly seem to be perpendicular to individual slope position on rock outcrop. Majority of the fans are being held in a parabolic filtration fan. Most individuals, however, have extended between 10 and 20 adjacent arms straight outwards from the fan at around ~ 45 degrees. All extended arms are on the far side of the fan as oriented by the video. It is possible these arms are spreading in response to stronger current further from the substrate. Current overall is extremely weak (2–3 cm/s), and outstretched arms are positioned with the food grove downstream from current. Differing filters have been discussed as having a range of successful function based on current conditions (Loudon and Alstad, 1990; Baumiller 1997) and that dense filters, as with the many armed *Endoxocrinus*, may have limits on feeding ability in low currents (Baumiller 1988; Baumiller 1997). Under normal parabolic fan feeding position, *Endoxocrinus* may feed less effectively in very low velocities. The effect of repositioning and spreading these selected arms may be in creating a fan filter with lower density. Thus, this change in feeding posture may allow crinoids with a normally high density crown to continue feeding or increase feeding effectiveness in low velocity environments.

## Acknowledgments

The underwater time-lapse camera, flash, and control unit for Video 1 was designed and constructed by C. Arneson, Scripps Institute of Oceanography, and kindly provided by N.D. Holland, Scripps. We thank Prof. Tomasz Baumiller and Prof. Christopher Harley for their support with all *Florometra* mesocosm work. This project was supported by the MEOPAR and NSERC CREATE Training Our Future Ocean Leaders Program postdoctoral fellowships to A. Stevenson. We also thank Dr. Tomasz Baumiller and Chris Byrne for support with video or still recording during submersible dives in Roatán along with our submersible pilot, Karl Stanley.

# References

Baumiller, T. K. (1988). Effects of filter porosity and shape on fluid flux: implicationsfor the biology and evolutionary history of stalked crinoids, in R.D. Burke, P.V. Mladenov, P. Lambert and R.L. Parsley, eds., Echinoderm biology: Proceedings of the Sixth International Conference, Victoria, 23–28 August 1987, 786 pp. A.A. Balkema. Rotterdam.

Baumiller, T. (1997). Crinoid functional morphology. *The Paleontological Society Papers*, **3**, 45–68.

Baumiller, T. K., LaBarbera, M., & Woodley, J. W. (1991). Ecology and functional morphology of the isocrinid Cenocrinus asterius (Linnaeus) (Echinodermata: Crinoidea): in situ and laboratory experiments and observations. *Bulletin of Marine Science*, **48**, 731–748.

Byrne, M., & Fontaine, A.R. (1983). Morphology and function of the tube feet of Florometra serratissima (Echinodermata: Crinoidea), *Zoomorphology*, **102** (1983), 175–187.

Lahaye, C. A., & Jangoux, M. (1985). Functional morphology of the podia and ambulacral grooves of the comatulid crinoid *Antedon bifida* (Echinodermata). *Marine Biology*, **86** (3), 307–318.

Loudon, C., & Alstad, D.N. (1990). Theoretical mechanics of particle capture: predictions for hydropsychid caddisfly distributional ecology, *American Naturalist*, **135**, 360–381.

McEdward, L. R., Carson, S. F., and Chia, F. S. (1988). Energetic content of eggs, larvae, and juveniles of *Florometra serratissima* and the implications for the evolution of crinoid life histories. *International Journal of Invertebrate Reproduction and Development*, **13**(1), 9–21.

Messing, C. G., Ausich, W.I ., & Meyer, D. L., Feeding and Arm Postures in Living and Fossil Crinoids. Treatise on Invertebrate Paleontology, Echinodermata, Crinoidea, Revised. Submitted.

Meyer, D. L (1982). Food composition and feeding behavior of sympatric species of comatulid crinoids from the Palau Islands (Western Pacific). In Lawrence, J.M., ed., *Proceedings of the International Conference, Tampa Bay*. Rotterdam: A. A. Balkema, pp. 43–49.

Meyer, D. L. (1997). Reef crinoids as current meters: feeding responses to variable flow. *Proceedings of the 8th International Coral Reef Symposium*, **2**, 1127–1130.

Meyer, D. L., LaHaye, C. A, Holland, N. D., Arneson, A. C., & Strickler, J. R. (1984). Time-lapse cinematography of feather stars (Echinodermata:

Crinoidea) on the Great Barrier Reef: demonstrations of posture changes, locomotion, spawning, and possible predation by fish. *Marine Biology*, **78**,179–194.

Mladenov, P. V., and Chia, F. S. (1983). Development, settling behaviour, metamorphosis and pentacrinoid feeding and growth of the feather star Florometra serratissima. *Marine Biology*, **73**, 309–323.

Taylor, K. H., Rouse, G. W., & Messing, C. G. (2017) Systematics of *Himerometra* (Echinodermata: Crinoidea: Himerometridae) based on morphology and molecular data. *Zoological Journal of the Linnean Society*, **181**, 342–356.

# Cambridge Elements ☰

## Elements of Paleontology

### Editor-in-Chief
Colin D. Sumrall
*University of Tennessee*

### About the Series
The Elements of Paleontology series is a publishing collaboration between the Paleontological Society and Cambridge University Press. The series covers the full spectrum of topics in paleontology and paleobiology, and related topics in the Earth and life sciences of interest to students and researchers of paleontology.

**The Paleontological Society** is an international nonprofit organization devoted exclusively to the science of paleontology: invertebrate and vertebrate paleontology, micropaleontology, and paleobotany. The Society's mission is to advance the study of the fossil record through scientific research, education, and advocacy. Its vision is to be a leading global advocate for understanding life's history and evolution. The Society has several membership categories, including regular, amateur/avocational, student, and retired. Members, representing some 40 countries, include professional paleontologists, academicians, science editors, Earth science teachers, museum specialists, undergraduate and graduate students, postdoctoral scholars, and amateur/avocational paleontologists.

**Paleontological**
S  O  C  I  E  T  Y

## Cambridge Elements ≡

# Elements of Paleontology

Printed in the United States
by Baker & Taylor Publisher Services